Animals Live Here

# Life in the Arctic

By Connor Stratton



level 2 little blue readers

www.littlebluehousebooks.com



Little Blue House is distributed by North Star Editions:
sales@northstareditions.com | 888-417-0195

Produced for Little Blue House by Red Line Editorial.

Photographs ©: tryton2011/Shutterstock Images, cover; FloridaStock/Shutterstock Images, 4; rpbirdman/iStockphoto, 7 (top); Menno Schaefer/Shutterstock Images, 7 (bottom), 24 (bottom left); Jamen Percy/Shutterstock Images, 9, 24 (top left); Artpilot/iStockphoto, 10–11; Giedriius/Shutterstock Images, 12; 4FR/iStockphoto, 15 (top); Fitawoman/Shutterstock Images, 15 (bottom); sirtravelalot/Shutterstock Images, 17; Mikhail Cheremkin/Shutterstock Images, 18, 24 (bottom right); cmeder/iStockphoto, 20–21; vladsilver/Shutterstock Images, 23, 24 (top right)

**Library of Congress Control Number: 2019908251**

**ISBN**
978-1-64619-019-5 (hardcover)
978-1-64619-058-4 (paperback)
978-1-64619-097-3 (ebook pdf)
978-1-64619-136-9 (hosted ebook)

Printed in the United States of America
Mankato, MN
012020

## About the Author

Connor Stratton enjoys exploring new places, spotting new animals, and writing books for children. He lives in Minnesota.

# Table of Contents

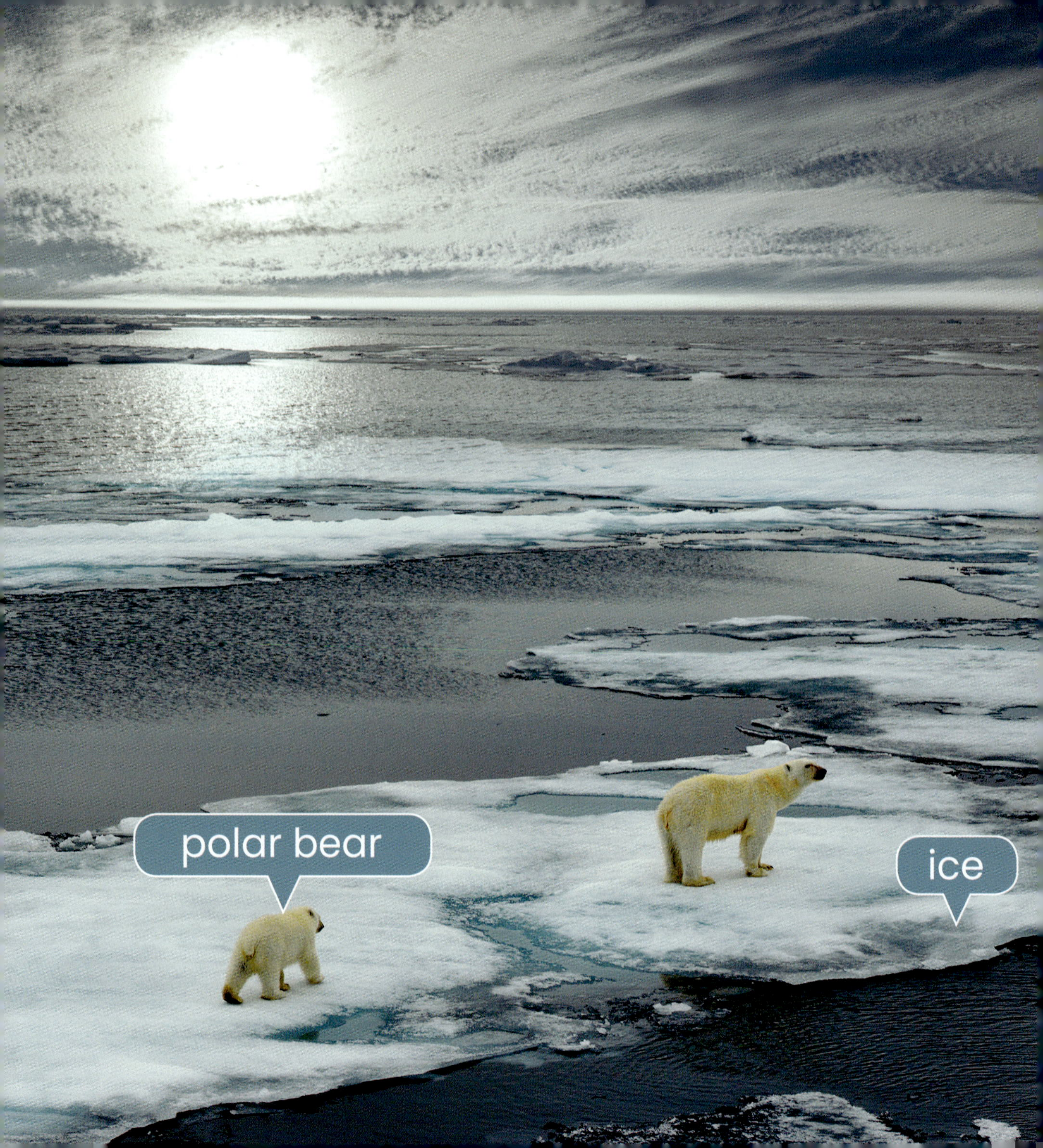
polar bear
ice

# Arctic Animals

The Arctic is a cold place filled with snow and ice. Polar bears live in the Arctic. Many other animals live here too.

Owls live in the Arctic.

These birds have white feathers.

Puffins live in the Arctic.

These birds have big beaks and webbed feet.

owl
feather

beak
puffin
feet

Lynx live in the Arctic. These large cats hunt at night.

lynx

Caribou live in the Arctic. They have antlers on their heads. Caribou are also called reindeer.

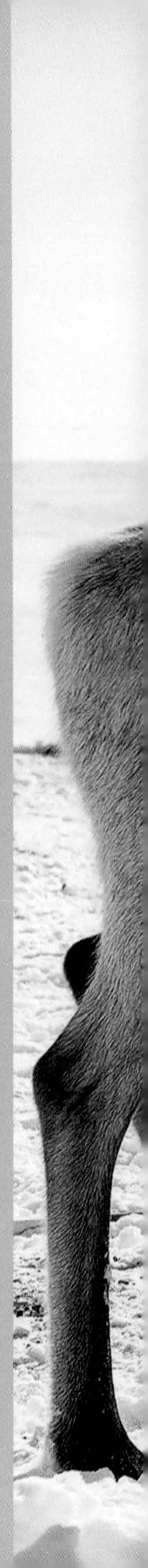

antler
caribou

musk ox
fur

# Warm Fur

Musk oxen live in the Arctic. They have thick fur coats. Their fur keeps them warm.

Wolves live in the Arctic.
They have sharp teeth.
Foxes live in the Arctic.
Their tails keep them
warm when they sleep.

teeth
wolf

fox
tail

Hares live in the Arctic.
Their fur turns brown
when the snow melts.

hare

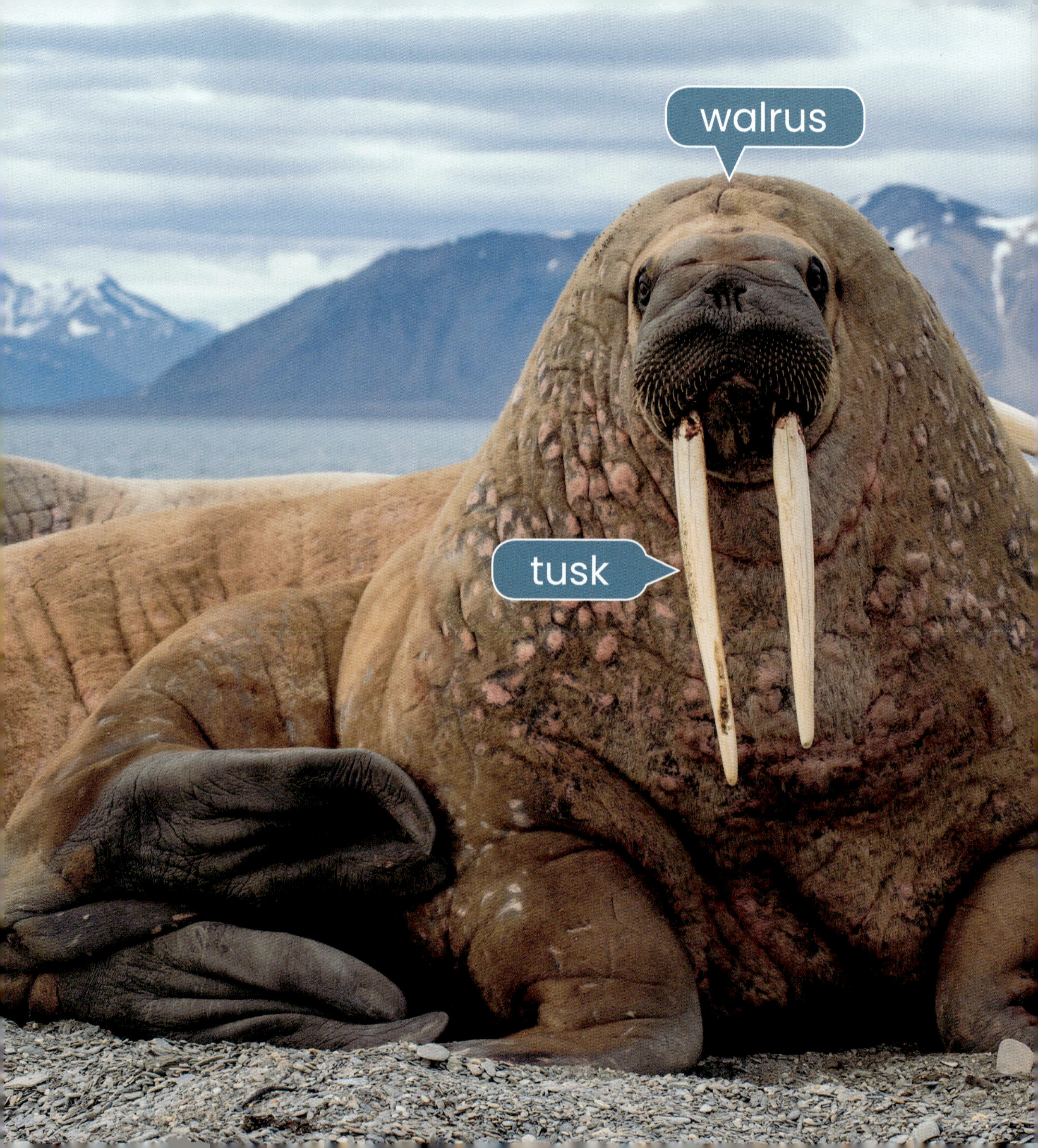
walrus
tusk

# Cold Waters

Walruses live in the Arctic. They have large tusks. They can swim in very cold water.

Whales live in the Arctic.
They have white skin.
They are the biggest
animals in the Arctic.

whale

Seals live in the Arctic.
They can dive into
the water.
They use their flippers
to swim.

seal
flipper

# Glossary

lynx

seal

puffins

walruses

# Index